Wildlife
in Papua
New Guinea

A TRIBUTE

To Paulis Arek.

who died while this book was in production,

ACKNOWLEDGEMENTS.

Many of my friends will recognise these photographs, ranging from the very rare Hawk Owl in the bird section to the various mammals, snakes, orchids etc. they have helped me photograph. To these people may I here say, *thank you*.

Special thanks also to Brian Coates of Port Moresby for allowing me to use, in particular, his bird studies.

This edition produced
for
Robert Brown and Associates
Port Moresby, New Guinea.

Wildlife in Papua New Guinea

Eric Lindgren

Drawings : Vali Herzer

Golden Press

Plate 2. Background illustration on title page, White-tailed Rat.

3

First published 1975 by Golden Press Pty. Ltd.,
Sydney, Australia.
© Peregrine Photographics
ISBN 0 85558 503 X
Printed in Hong Kong by Toppan Printing Company.

CONTENTS

INTRODUCTION

4

INTRODUCTION

The island of New Guinea is a meeting place for the animals and plants from the continent of Asia, from the islands to the north and west, and from Australia to the south. But paradoxically, at the same time as being a meeting place, New Guinea also separates Asian and Australian wildlife. Reasons can be sought in the habitats the animals occupy, habitats made up of plants from the western land masses, plants from Australia, and plants which evolved in New Guinea.

For millions of years the Australian continental plate, which includes the vast southern lowlands of Papua, has been drifting northwards. Becoming drier, it was covered with desert-adapted *Acacia* and *Eucalyptus* trees and fringed by wetter coastal forests. Early Australia carried a rich marsupial fauna. The slow drift northwards was interrupted as the continental plate butted up against a northern submerged land mass. Along the line of contact a slow buckling occurred and the central mountain chain of present New Guinea emerged. Heavy rainfall and volcanic activity, created from the disturbance to the earth's crust, resulted in a rapidly changing environment. As the mountains grew, new high altitude vegetation associations developed from the cold-loving species of Australian plants and invaders from the jungles to the west. New habitats were formed. With the creation of troughs between Australia and the new land mass, the common fauna divided to become two distinct units clearly related. One, a suite of marsupials, rodents and bats adapted to high rainfall, high temperatures and closed canopy forests, and the other adapted to an increasingly arid, open woodland.

New Guinea now provides a rich variety of habitats for its animals: from reef and swampland at sea-level, through swamp forest and lowland rainforest on the coastal plains and foothills, to mid-mountain beech and pine forests and, at the highest altitude, stunted moss forest ("elfin woodland") and alpine grassland. Highest peaks on the island rise over 5000 metres. Mount Carstenz of Irian Jaya and Mt. Wilhelm of Papua New Guinea present eerie bare-rock meadows and slopes pervaded by an out-of-this-world atmosphere. Permanent snow on Carstenz Top, near the equator, is reminiscent of the majesty of the African Mt. Kilimanjaro. It is here that many of the plants and animals reaching lowland New Guinea in past cold climates now occur.

Orchid-laden tree ferns, unique to the island, intermingle with heath-like plants related to those of southern Australia and Tasmania. Lower down the mountain slopes are other southern plants. The Antarctic Beech *Nothofagus* lives here. There are fifteen species in New Guinea, and it is also found in New Zealand, Patagonia, Tasmania, Victoria, the New England plateau of Australia, and, surprisingly, New Caledonia.

Bound up so closely with the invaders from the south are the tropical Asian plants—rainforest trees, shrubs, orchids—and intimately involved with the migrants is the richly endowed flora which evolved in New Guinea. It is a vast flora, ten thousand species of plants in 810 000 square kilometres. Australia by comparison has twelve thousand species in nearly eight million square kilometres.

Within New Guinea have evolved 2 700 species of orchids, numerous *Rhododendron* and a bewildering variety of forest giants at all altitudes. The multitude of plant species reflects the multitude of habitats available for the animals of the island to occupy.

The mosaic of mountains, valleys, major river systems, and the complex of plants, provide the framework which determined the present day wildlife of New Guinea.

New Guinea is a meeting place. The eastward-moving wildlife of Asia, from humid tropical forests, found the hot wet forests to its liking. The animals of the southern savannahs and recent invaders from Australia found the seasonally dry *Eucalyptus* and *Melaleuca* woodlands to their liking.

New Guinea is a barrier. Southern savannahs bar the way to migrants from humid rainforest. Humid rainforest bars the way to dry-country Australian animals.

This is the paradox of New Guinea.

This book shows some of the wildlife of New Guinea.

I hope you enjoy it.

5

MAMMALS

New Guinea, Australia and South America are the main centres of distribution of marsupials. Each area has its own distinctive species, reflecting the evolutionary pathway taken by the stocks on each land mass when they separated in the past and slowly drifted apart.

In New Guinea, as elsewhere, the marsupials were later joined by groups of mammals from nearby land surfaces. Two such groups—the rodents and the bats—managed to colonise the island.

Papua New Guinea now houses a mixture of endemic species of marsupials, rodents and bats and recent Australian invaders of these groups. In the rainforest of the lowlands, the endemic *Dorcopsis* wallabies shelter beneath the canopy. Nearby in the *kunai* grassland, the Australian Agile Wallaby grazes. Similarly the endemic Raffray's Bandicoot selects a different habitat from the Australian Short-nosed Bandicoot.

As might be expected in this land of high rainfall and deep valleys, the swiftly rushing streams house a suite of "water rats", related to their Australian cousins. In the highlands the endemic silky-furred *Crossomys* seeks its food in the turbulent streams; in the southern lowlands the Australian Water Rat occupies the slowly meandering rivers of the plains. Some recent rodent invaders are identical to their Australian counterparts—the Dusky Field Rat, *R. sordidus* for example.

In the air above, the pattern is repeated. Giant flying foxes and tiny insectivorous bats include species of both New Guinean and Australian origin. Included also are small numbers of colonisers from Indonesia, for bats can cross the water barriers which hold back the terrestrial mammals.

MAMMALS—190 species, 11 families

6

One night, plunging into the forest two hours after the sun had set, I catch sight of two bright eyes staring unblinkingly at me. Are they in a tree, on a branch? On the trunk? No, for they are at the bottom of the animal—a bat! Yellow spotted wings, large head and long tube nostrils identify this as a fruit eater. The pupils contract as I come closer. Then off it goes for I must have startled it. Farther into the forest, I see another pair of eyes. The same fiery hue and it is another Tube-nosed Bat. It is certainly a fruit-eater, for while hanging upside-down, it is using its belly as a table. There on the short fur is the half-eaten fruit of a rainforest tree. So tame is this bat that it waits while each camera flash makes the pupils smaller and smaller until only pinpoints remain. Overhead I hear the clumsy leathery wing flapping of a flying fox, a relative of this small tube-nose.

To see the rainforest mammals you must meet them in their world. In daytime the large species are active; the Agile Wallaby of the grassland feeds on the wide river plains; the Rusa Deer stands and watches as your boat passes by. But at night—and the darker the better—you will see all those which hide away during the day. The beam of my head torch catches the fiery orange-red light in their eyes. The mammals are out. Low to the ground, bright lights twist rapidly as the owner strives to catch my scent. Move closer, quietly now. Then I have it; it is Govei the silky grey wallaby of the rainforest.

Another night, two small headlights seem to run across the ground. It is only a rat, probably a White-footed Rat. More interesting is the movement of a ringtail in the trees. Walking slowly, it moves closer. Some leaves are in the way. Move left. Yes, it is a ringtail, but not the ringtail of the Australian temperate rainforests; this is a Coppery Ringtail and a young one. Distracted by a bright flickering to my right—I glance at the sharp white pinpoint of a firefly. The Ringtail vanishes.

At night, the savannah paths will reveal to you the bandicoots, rats, wallabies and, with luck, a Spotted Cuscus. The rainforest will delight you with those I have seen, and the be-whiskered Tree-Rat, or the friendly Mosaic-tailed Rat. Higher up perhaps will be a Tree Kangaroo, a Ground Cuscus or even a Striped Phalanger. The rainforest at night is a world apart!

The **Long-beaked Echidna** (plate 6) is a large relative of the Australian Short-beaked Echidna. Found in mid and high-mountain forests between 1000 and 4000 metres its beak is specialised to feed mainly on earthworms from the ground litter.

The **Spotted Cuscus** (plate 7).

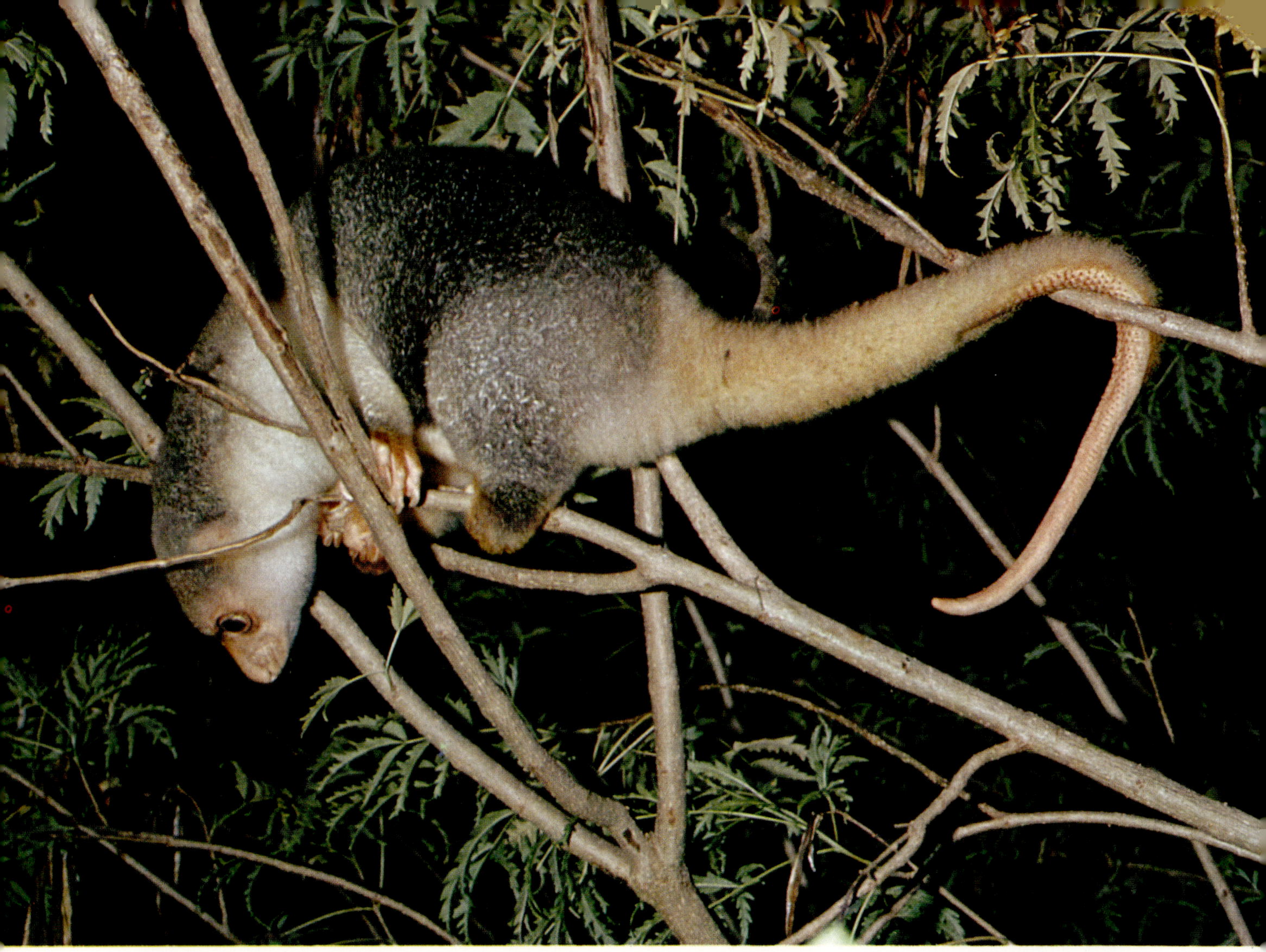

8

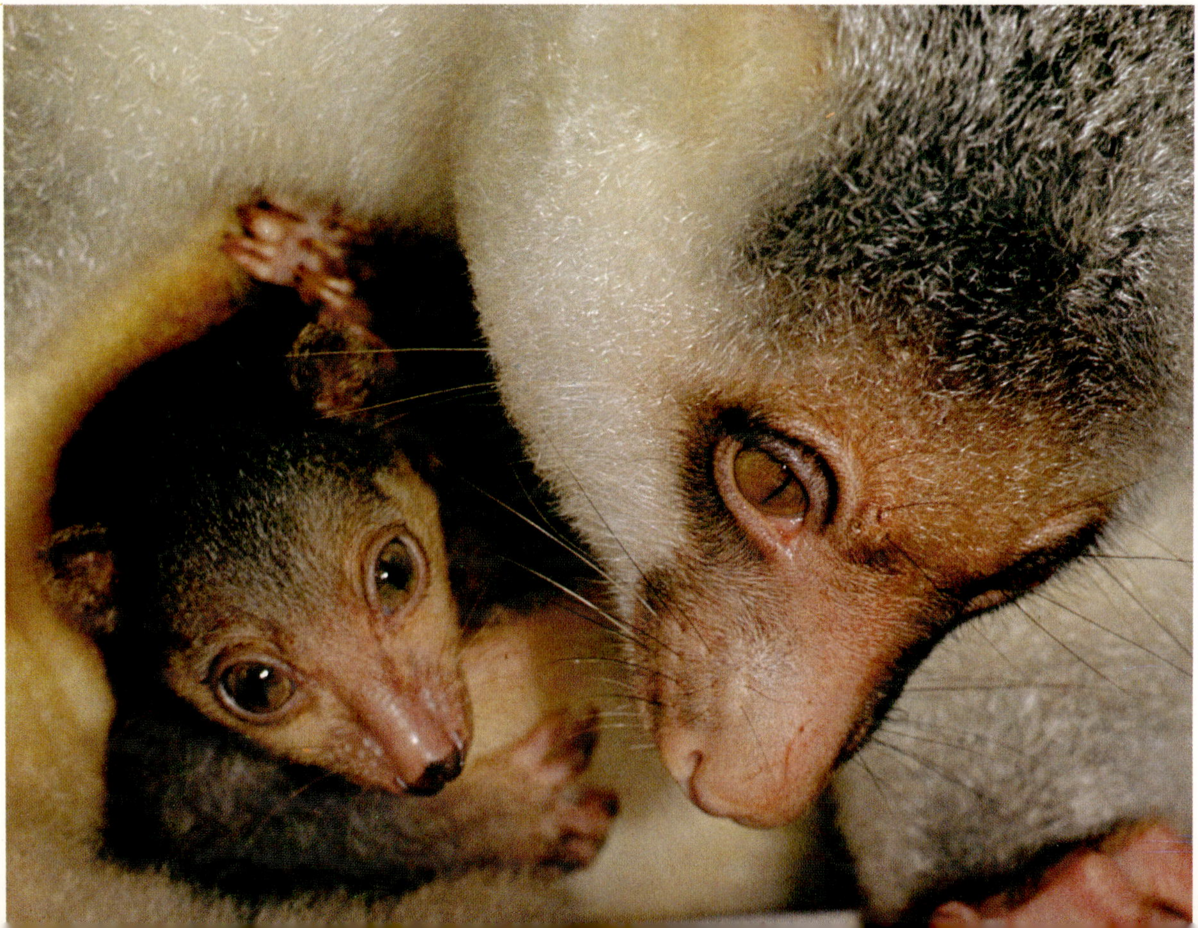

9

A member of the Possum family, the **Spotted Cuscus** (plates 7–11), is reputed to be slow and sloth-like. In reality it is a nocturnal fruit and leaf eater which punctuates periods of rest with active food-seeking. Males are generally smaller than females and have brown spots on the upperparts. Females lack the spots and vary in colour from orange-brown to silvery-grey according to the area they come from. Pure white animals occur in northern coastal areas around Madang and almost totally black individuals are found on Manus Island.

The **Sugar Glider** (plate 12) and **Painted Ringtail** (plate 14) are both tree-dwelling relatives of the Spotted Cuscus. Sugar Gliders are a wide ranging species occurring from the forests of coastal eastern and northern Australia into the lowlands of Papua New Guinea. They rarely occur above 1 500 metres and prefer humid forests rather than the cooler and wetter mountain forests. Social groups of up to ten individuals frequently build leaf nests in the crown of coconut palms and maintain contact by a system of characteristic calls and scents.

White-tailed Rats (plate 13) are perhaps the largest species of rat in Papua New Guinea. Fully grown individuals may reach the size of a small pussycat. They are common in the rainforests from sea-level to 2 000 metres and forage equally as well on the ground as in trees. Rather tame, these rats will approach a quiet observer at night and show little fear. Another ground-dwelling species, **Raffray's Bandicoot** (plate 15), seeks fruit and insects on forest floors to 3 000 metres. The long snout and strong claws of the front legs reflect the feverish grubbing activity so typical of all bandicoots. Seeking insects and roots beneath the soil, they leave areas of characteristic conical diggings where food has been plentiful. As the young mature they lose the bright rufous colouration in the picture.

15

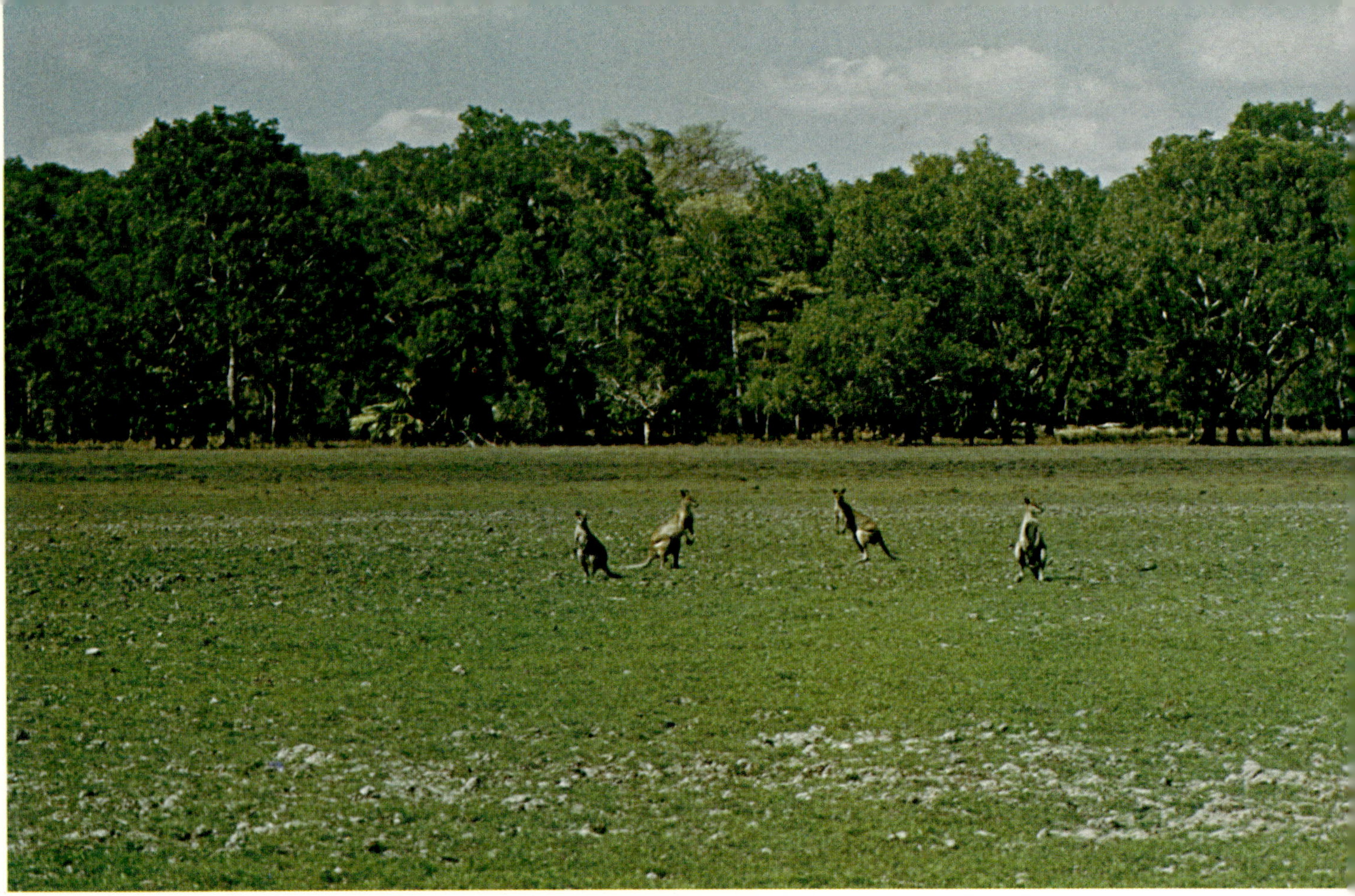

17

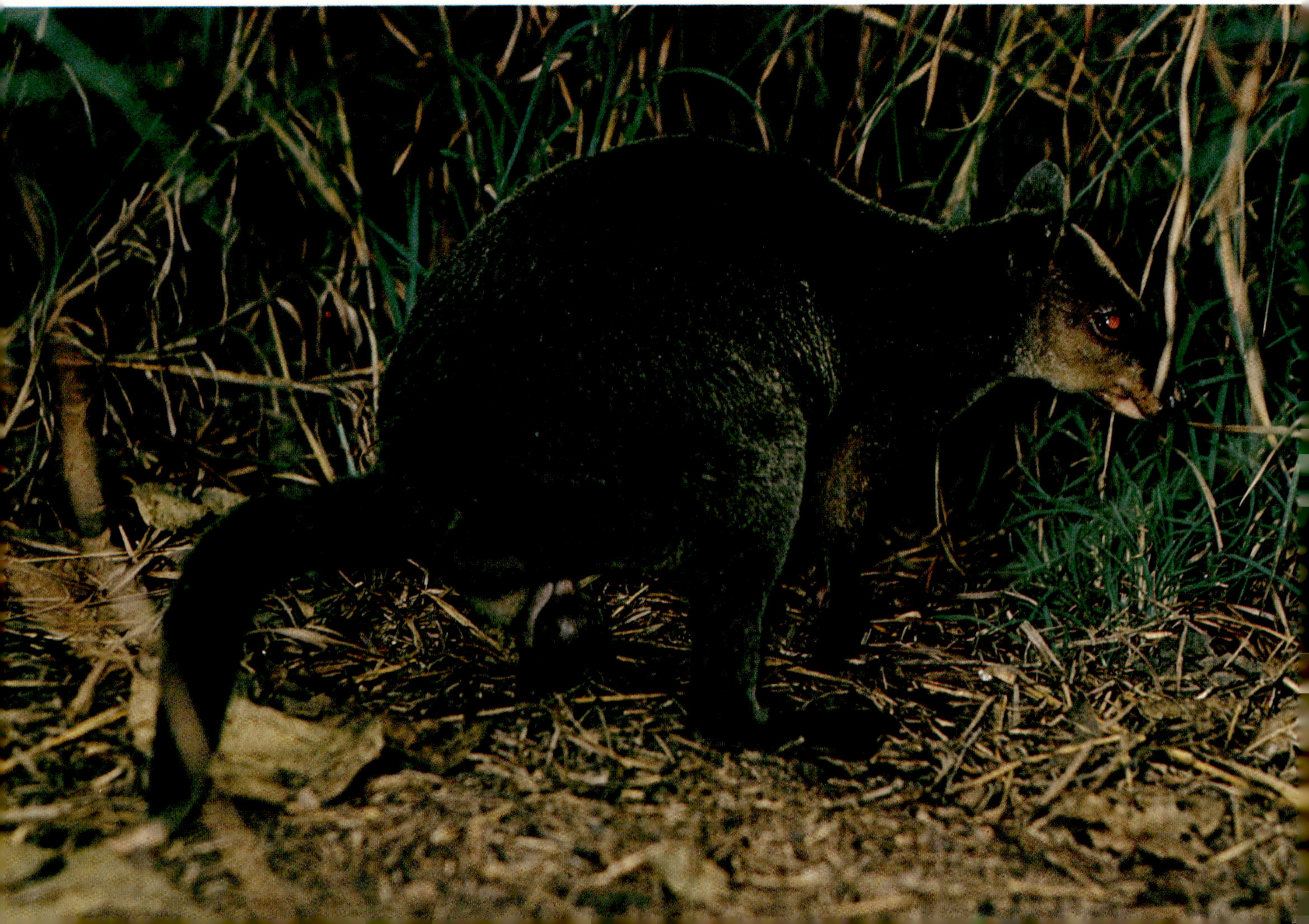

Unlike Australia, where kangaroos and their relatives have developed into a rich complex of open ground species, Papua New Guinea shows almost equal diversity of ground-dwelling and tree-dwelling forms. The **Agile Wallaby** (plate 16) of the Papuan savannahs occurs also in northern Australia. This is the *Magani* of the Motu-speaking Papuan peoples. In contrast the **Forest Wallaby** (plate 17) occurs in dense lowland rainforests and is known as *Govei* to Motuans. The photograph shows the characteristic tail-tip stance of New Guinea forest wallabies, species which range from sea level to 3 000 metres.

18

Matschie's Tree Kangaroo (plate 18), one of four such species in Papua New Guinea, is similar to the brightly coloured Goodfellow's Tree Kangaroo from the central mountain forests. However it lacks the distinctive barred tail of the latter species and is confined to the Huon Peninsula to the north of Lae.

20   The **Pen-Tailed Possum** (plates 19 and 20) is a small mouse-sized possum which has an unusual fringe of long hairs down each side of the tail. Flattened tails are characteristic of many species of gliding mammals and it is puzzling in this species, which does not glide. Pen-tailed Possums live in the shrub-layer of lowland rainforest up to 2 000 metres.

Deer do not occur naturally in Papua New Guinea. **Javan Rusa** (plate 21) were introduced to a number of localities during the first half of this century. It is a swamp and grassland species from certain islands in Indonesia and prefers similar habitat in New Guinea. The largest numbers are found in the vast wetlands south and east of the Fly River in Papua. A second species of deer, the Axis, occurs in small numbers on the outskirts of Madang.

21

Bats and rodents are the two groups of Papua New Guinean mammals with the largest numbers of species. The **Horseshoe Bat** (plate 22) is an insectivorous bat which uses echo-sounding for locating and capturing its prey in flight, and for finding its way in the caves and tunnels where it roosts. The broadly flattened nose, shaped like a horseshoe, is most probably a complex reflector or receptor of the high-pitched sounds used in navigation. The female Horseshoe Bat in the photograph clasps a baby within its wing membranes. It is from a maternity group of bats in an abandoned copper mine near Port Moresby.

22

23

Although insect-eating bats are, in general, relatively small species, and fruit bats or flying foxes are larger, often with a wing span to two metres, the **Tube-nosed Bat** (plate 24) and **Black-bellied Fruit Bat** (plate 23) are both small fruit eaters of the lowland forests. Both are colourful: the black belly and orange back of the Black-bellied Fruit Bat distinguish it from all other New Guinea bats. The tube nostrils and irregular blotching on the wing membrane separate out two similar genera of tube-nosed bats from other fruit bats. The bat shown at right is using its belly as a table to hold the rainforest fruit it is eating.

BIRDS

New Guinea birds, with their brilliant colour, surpass those of any other country. Nothing transcends the beauty of a filmy-plumed bird of paradise, *Paradisaea*, at the height of its display. Brilliant yellow plumes of the Lesser Bird of Paradise are thrown above the body to stand erect like the plumes of a Roman centurion's helmet. Frenzied calls attract a mate who must choose from the dozen or more males so determined in their display. The pattern is repeated in the red-plumed *raggiana* species and in the incredible dances of the Flag-birds and the Superb bird.

In the past, each new species was proclaimed with a royal title: Count Raggi, Princess Stephanie, Emperor of Germany, Queen Carola, King Bird of Paradise. Magnificent! Superb! In life, the birds reflect the proclamation, for their displays are unique to the world of birds. The filmy-plumed group select a display tree and show off in the early morning and evening. The King Bird displays on its own, on a perch in a shaft of sunlight, puffing itself up and pivoting with the green jewels on its tail tips dancing above its head. The Superb is aggressive, erecting the ruff around its neck and whip-cracking at the female while performing its dance.

Many of the smaller birds of New Guinea are brilliant. The Black Sunbird, sombre in the shade, reveals shining patches of iridescent green, purple and blue in the sun. Like the humming birds of South America, this species is a living jewel and is even able to hum. The golden *Pachycare*, related to the Whistlers of Australia, is also extremely colourful, as is the Blue Wren Warbler, the Yellow-backed Whistler and the Papuan Parrot Finch.

BIRDS—650 species, 75 families

25

Three thousand five hundred metres up Mt. Albert Edward, I was in a cold, stunted landscape and there before me was one of the most beautiful birds I have ever seen. It was smallish, subtly coloured in a mixture of dove-greys, blues and yellows, and had a black head. There was none of the bold bright brassiness of the lowland parrots, nor the brilliance of the Gouldian Finch of Australia. Rather it was the blending of soft pastel shades, smooth contours and the impression of an exquisite finish that identified it as the Crested Berrypecker.

This was a surprising landscape, for two of the first birds I had seen were familiar lowland species. From beneath my foot, a whistle and the sudden whirring of wings disclosed a Brown Quail. Then from the grass came a familiar whistle which I could not place. When it was repeated, I recalled it. I had last heard the Tawny Grassbird on the banks of the Bensbach River away in the remote southwest of Papua.

Papua surprises me. In the vast Eucalyptus savannahs of the Western District, here are the familiar birds—a Magpie from Australia, the Blue-winged Kookaburra, Red-winged Parrots and White-breasted Sea Eagles. Here too are the Island birds—the Greater Streaked Lory with short wings and stubby tail, the Greater Bird of Paradise, apoda, an immature male, shy and without plumes, the bright buzzing Black Sunbird flashing its jewelled throat and head.

Papua New Guinea presents these puzzles. The unexpected in one place—Olsobip and an Osprey four hundred kilometres from the sea, three hundred metres above sea level near the central Star Mountains of the border. Papua New Guinea presents a challenge—beneath the unbroken canopy of the rainforest are a multitude of calls. Brief silhouettes give clues to their origin. Then suddenly a tree explodes and a dozen fruit pigeons wing off to settle and disappear. Here, in the lowland forests close to Port Moresby, with patience you will see the blood-red Little King, hear the raucous kwarking of Raggiana, discover the incredibly long nest of the Rufous Babbler and wonder at the tiny Pigmy Parrots.

With patience they will be yours to view.

Mid-mountain forests of the central cordillera, 1 400 to 2 600 metres above sea-level, are the home of the **Black Sicklebill Bird Of Paradise** (plate 25). In full display the breast plumes are spread to meet above the head and the mouth is opened widely to display the bright yellow lining. The loud whip-call of this species penetrates the forest up to a kilometre.

Common Paradise Kingfisher (plate 26)

The twenty-four species of New Guinea king-fishers range in size from the tiny **Little King-fisher** (plate 27), 100 mm in length, to the giant Blue-winged Kookaburra, 400 mm in length. A lowland forest bird, the Little Kingfisher prefers the vicinity of water where it catches fish and insects. Individuals habitually sleep on the same twig for a number of nights and the bird photographed was regularly found at its roost over a period of two weeks. The **Hook-billed Kingfisher** (plate 28) is also a lowland forest species, to 1 300 metres, which appears to be nocturnal in habit as its slow whistling notes are frequently heard at night. It is reputed to dig in the muddy soils of the rainforest in search of food.

Unlike the **River Kingfisher** (plate 30) which ranges from Europe, through Asia and New Guinea, to the Solomon Islands, the **Rufous-bellied Kingfisher** (plate 29) is restricted to New Guinea and nearby islands. As its name suggests, the River Kingfisher always associates with water, both inland fresh waters and wooded ocean beaches. It feeds upon fish and water insects. The Rufous-bellied Kingfisher is reminiscent of a kooka-burra. Its flight is faster but the call has the same quality. A lowland forest species, to 500 metres, this bird feeds on a variety of insects and small vertebrates.

27

28

29

30

31

32

Parrots in New Guinea number forty-six species. The small **Buffy-faced Pygmy Parrot** (plate 31), about 75 mm, is one of the smallest of the world's parrots, shown here half-again larger than life size. These birds nest in an arboreal termite mound, as do many parrots and kingfishers, and out of the breeding season groups of birds sleep in the same nest cavities. The food eaten by these tiny parrots has not been accurately determined. They restlessly explore crevices and tear off flakes of rotting bark from the trunks of trees in the manner of tree-creepers. It has been suggested that they feed on fungus and algae on the bark. **Goldie's Lorikeet** (plate 32) is one of a group of brush-tongued parrots which seek nectar and pollen from a variety of rainforest trees. It is similar to the Varied Lorikeet of northern Australia and like that species has no red cap in the juvenile plumage, as pictured.

33

34

The largest group of lorikeets in New Guinea, the eight *Charmosyna* species, are social birds moving about in small flocks. The **Fairy Lory** of high mountain forests, 1 600 to 3 000 metres, occurs in a "normal" red form (plate 33) at the lower parts of its range. At higher altitudes, a dark form occurs in which most of the red is replaced by black (plate 34). Both forms have the elongated yellow-tipped central tail feathers characteristic of the species. Similar in pattern to the Fairy Lory, the widespread lowland forest **Black-capped Lory** (plate 35) is a heavy-bodied pugnacious species which tends to be solitary except at roosting time when assemblies of scores of birds may congregate in the one tree to sleep. These disperse in the early morning as individuals go their own way to feed. Flowers, fruit and nectar form their diet.

35

A bird of the sky and frequenter of water-side habitats, the **White-breasted Sea Eagle** (plate 36), ranges from India through Malaysia to New Guinea and Australia. It is the largest eagle of the lowlands but as the occupant of the top of a food-chain is not frequently seen. Young birds are brown, more lightly speckled, and may be mistaken for the rare Wedge-tailed Eagle, which occurs only in the Western District of Papua. The shorter, whitish tail of the sea eagle distinguishes it.

37 Broad-winged hawks form a confusing group in Papua New Guinea as they are not frequently seen and, particularly in juvenile plumage, tend to look alike. This juvenile **Grey-headed Goshawk** (plate 37) is a lowland forest species, to 1100 metres, confined to New Guinea and adjacent islands. It loses the dark striping underneath as it matures and becomes a trim grey and white bird with a striking orange face and legs.

Like the birds of paradise and parrots, pigeons form an important group of birds in Papua New Guinea. There are forty-four species on the mainland. The brightly coloured fruit-doves are chiefly lowland forest species which congregate on fruiting trees to feed. The **Little Coronetted Fruit Dove** (plate 39) shows the typical pattern of green body and bright crown and underparts. **The Nicobar Pigeon** (plate 38) is found chiefly on small islands along the northern coast of New Guinea and from Malaysia east to the Solomons. The male has an all-white tail, the female's is tipped with black. A dainty inhabitant of the forest floor, to 1 700 metres, the **Rufous-throated Ground Dove** (plate 40) is a nervous species, rarely seen unless the observer sits quietly and waits for one to pass by.

41

42

43

Birds of the night—nightjars, frogmouths and owls—are common in the forests and savannahs of Papua New Guinea. It is unusual not to hear the monotonous *chop . . chop . . chop* of the **White-tailed Nightjar** (plate 41) throughout the lowlands and mid-mountain forest, particularly in disturbed areas. Displaying conspicuous white patches near the wing tips, this species is easily recognised when flushed from its roost on the ground. In contrast the ghostly moth-like flight of the small **Bennett's Owlet-Nightjar** (plate 43) lacks distinguishing features. This species, however, is the only owlet-nightjar to be expected from the majority of lowland forest areas. Both the **Papuan Frogmouth** (plate 42) and the **Little Frogmouth** (plate 44) have an enormous beak like a frog's mouth. These species hunt at night, feeding principally upon ground-living insects and small verte-brates which they locate from a convenient perch. The downy Papuan Frogmouth chick pictured will grow to a long-tailed adult with a striking blood-red eye.

44

Both the Barn Owl family and the Hawk Owl **45**
family are represented in Papua New Guinea.
The former are typified by a distinctive heart-
shaped face ringed by a border of stiff feathers.
The latter lack this feature and are more hawk-
like in appearance.

A bird of the darker forests, the **Sooty Owl**
(plate 46) blends in well with its habitat. The
large eyes, larger than those of the **Grass Owl**
(plates 47 and 48) and the Barn Owl, indicate
the low light prevailing beneath the rainforest
canopy at night. The Grass Owl, as its name
suggests, is a bird of the open country, nesting
on the ground particularly in the wide grassy
valleys of the central highlands. The downy
chick (plate 48) is the same bird as pictured in
plate 47.

46 The **Papuan Hawk Owl** (plate 45) is a
distinctively long-tailed small owl of forests
from sea-level to 1 300 metres. It is a little-
known species in which the heavily streaked
underparts contrast with the barred upperparts. In the Port Moresby area and on the
D'Entrecastaux Archipelago, the **Brown Owl**
(plate 49) has streaked underparts similar to
the familiar Boobook Owl of Australia. Over
the greater part of its range, however, it is a
handsome bird with a sooty head and chestnut
breast, as pictured. The yellow eyes and a uniform blackish head seem to typify the species.

50

The deep litter of the floor of the rainforest is the habitat of a variety of ground birds, especially the litter of the humid, wet forests which suffer no dry season. Rotting leaf-fall, abundant fungus and sub-surface earthworms soon add to the soil. Three species of Eupetes, related to the babblers from India through to Australia exploit the animals of the litter. **Lowlands Eupetes** (plate 51) are found in the forest up to 600 metres. The female differs from the male either in having a white eyebrow, as pictured, or in having extensive brown and chestnut in the plumage. At a higher altitude from 600 to 1 400 metres, the **Mid-Mountain Eupetes** (plate 50) replaces the lowland species. It differs chiefly in the rich chestnut crown and back. Higher still occurs the High Mountain Eupetes, with black wing coverts spotted white.

51

Although the Quail Thrush are a group of dry-country dwellers in Australia, the one representative in New Guinea is found in lowland rainforests. This, the **Ajax Quail Thrush** (plate 52), has the same simple high-pitched whistle as the Australian birds.

The **Lowland Mouse Babbler** (plate 53), found from sea-level to 1 600 metres, is replaced by the Mid-mountain Mouse Babbler to 1 900 metres. This in turn gives way to the Mountain Mouse Babbler to 4 000 metres. These birds are all shy loud-voiced species which are easily overlooked unless the calls are known.

53

54

Above the forest litter and beneath the canopy, in the sparser leafy growth of small trees and shrubs, there lives a variety of birds. The second largest bird family in New Guinea, the Fly-catchers, with forty-nine species, occurs mainly here. These birds range from such familiar Australian species as the Willy Wagtail, through species similar to their Australian relatives such as the Monarchs, to distinctively New Guinean birds such as the two species of *Peltops*.

55

Fantails, twelve species, and Monarchs, eleven species, are the most frequently seen or heard flycatchers. The **Rufous-backed Fantail** (plate 57) occurs up to 1 000 metres and prefers the under-storey vegetation as its habitat. Although the **Sooty Thicket Fantail** (plate 56) occurs over the same altitudinal range it lives on the ground in palm thickets in swamp forest. Its loud protesting calls are most unlike those of other fantails.

The **Lowlands Flatbill** (plate 55) and **Olive Microeca** (plate 59) are both relatively quiet species of the undergrowth which sit watching for food, and dart out when an insect is seen. The **Spot-winged Monarch** (plate 58) and **Frill-necked Flycatcher** (plate 60) are more active and more vocal. Both are quick to scold an intruder with their harsh cries, but also have melodious whistling calls when undisturbed.

The **Blue Wren Warbler** (plate 54) is a large species of the undergrowth in lowland rainforest, allied to the Fairy Wrens of Australia, and thus in a different family from the above flycatchers.

56

57

58

59

60

61

The land of the birds of paradise: New Guinea. There are forty-three species which vary in size from the brilliant red **King Bird Of Paradise** (plate 61), a trim jewel with iridescent green circlets tipping the tail wires, to the long-tailed astrapias of the highlands. Typical to most people are the filmy-plumed *Paradisaea* species. The **Raggiana Bird of Paradise** (cover and plates 64 and 65) with its bright chestnut-red flank plumes is one of three widespread species. In the southern and western lowlands the Greater Bird of Paradise has dark-tipped coarse golden plumes. This species meets Raggiana in the Fly River area. Eastwards, south of the central cordillera, the red-plumed bird is a common and noisy inhabitant of the lowland rainforests. North of the Owen Stanley and Wharton Ranges, Raggiana continues in lowland forests to the Markham Valley. Here it is replaced by the soft-plumed Lesser Bird of Paradise along the northern coastal lowlands. This species has golden plumes tipped white. These are the common *Paradisaea* birds. Other more specialised filmy-plumed species occur in mountain forests or on islands.

62

63

Though the male birds of paradise sport brilliant colours, the females of each species generally have shades of brown and white. A characteristic female pattern, shown in the **Superb** (plate 62) and **Brown Sicklebill Birds of Paradise** (plate 63) is one of heavily barred underparts and uniform brown back. Only the female incubates the eggs and the sombre plumage is possibly of camouflage value.

Male **Raggiana Bird of Paradise** (plates 64 and 65).
In display the males of the *Paradisaea* species vigorously dance and dart about a cleared courtship area, throwing their wings above the body and reaching a climax with body horizontal, plumes erect and puffed over the back, wings drooped and shivering. Females, attracted to the males at the display court, choose a male and mate, and have little more contact while building, incubating and feeding the young.

The female **Magnificent Riflebird** (plate 66) exhibits a typical pattern, contrasting with the resplendent breast-shield of the male (plate 67). Both exhibit a bright lining to the mouth, a characteristic of many other species of birds of paradise and of value in display and communication.

68

The **White-eared Catbird** (plate 68) is common in the lowland rainforests to 1 000 metres. It is a large aggressive species, and though in the bowerbird family, it neither builds a bower nor has an elaborate display.

The **Rufous Babbler** (plate 69) **Rusty Pitohui** (plate 71) and **Grey-headed Whistler** (plate 73) are three lowland forest species which actively seek animal food. The Babbler and Pitohui occur only in New Guinea and are noisy social species which move through the canopy and under-storey in small restless flocks. The Whistler is a solitary bird.

Honeyeaters form the largest bird family in New Guinea, with sixty-five species. These three photographs indicate the variety in bill size from the **Long-billed Honeyeater** (plate 70) through the more typical bill size of the **Brown Xanthotis** (plate 72) to the atypical bill of the **Straight-billed Honeyeater** (plate 74). All are solitary species of lowland and mid-mountain forests.

69

70

71

72

73

74

REPTILES

Lizards and snakes are well represented in Papua New Guinea. There are about 170 species of lizards and 110 species of snakes.

Most abundant of the lizards are the smooth, shiny-scaled skinks. They range from the small ground and tree-dwelling species such as *Carlia* and *Emoia*, to the bizarre Giant Skink of Bougainville which grows to a metre in length.

Almost as common as the skinks, but less frequently seen because they hunt by night, are the soft velvety-skinned gekkos. Loria's Gekko, with its strikingly spotted back, represents one of the group of free-clawed species in which the toes are relatively thin and end in a claw. The Oceanic Gekko demonstrates the broad padded toes and tiny claws of the second group.

Dragon lizards are distinguished by the deep flap of skin beneath the throat, used to threaten an enemy, and the upright row of scales which form a crest on the neck and back.

The monitor lizards commonly known as goannas in Australia form the final family of lizards in Papua New Guinea.

Seven species of poisonous and dangerous front-fanged snakes are found mainly in the lowlands of Papua. Apart from the Death Adder these are swift, generally dark-coloured snakes which are active both day and night.

The other main families of snakes include the pythons, with many species of *Liasis* and the beautiful Green Python. The back-fanged snakes are tree-dwelling species seeking birds and tree frogs for food. These do have poison glands, but with grooved fangs at the back of their mouths they are relatively harmless.

REPTILES—280 species, 13 families

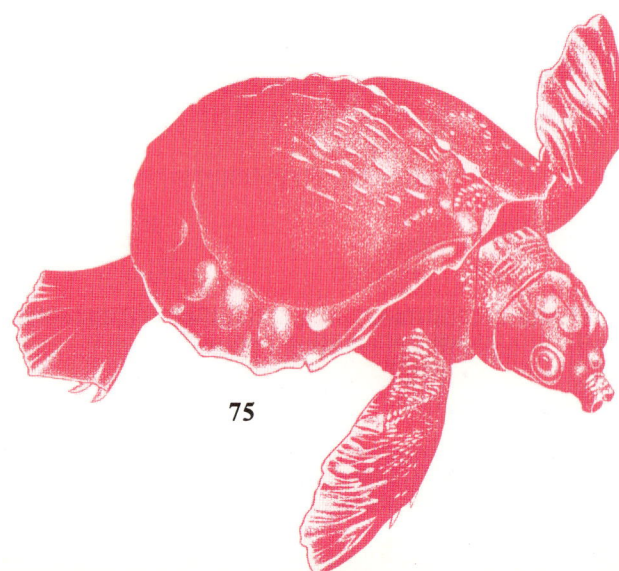

75

Tonight, in my favourite patch of rainforest, it has been a reptile night. Almost at once as I entered the forest I spotted small eyes shining at me from beneath the canopy. The small red eyes pressed close to the tree, too close for a mammal, too bright for a spider. Down the trunk they came and I saw the owner—a gekko. It was painted to match the bark. In the daylight I would have missed it. Here at night its eyes betray it. As I move closer, my light disturbs it and it jumps. It is two metres from the ground and it jumps for the tree nearby with legs splayed, claws ready to grip. It lands with a thump, catches hold and runs to the far side.

Moving quietly I catch it again with my light. Bright light shines in its eyes and again it jumps—this time to the ground and scampers off, scrambling over dead mossy logs. It pauses, licks its eye to clear it of soil, and I pounce. It is, strangely enough, a House Gekko, here in the forest. It is not the small pallid lizard you see on your wall but twenty centimetres of solid squirming individual. Be careful or you may damage its skin. Look at its toes to confirm your thoughts. The scales go right across the sole of each toe and act as a suction pad, holding each toe to the trunk. These pads will hold on the top, bottom or sides of a branch. Having identified it, I let it go. Up the tree it scampers until my light no longer shows me a lizard, only two red eyes.

Later, in my wanderings, I encounter another lizard. It comes sliding and slithering across the wet mud of the rain forest, like a fish out of water. It is shiny, with black and white striped sides. I have not seen this one before. Scampering, slipping, I hasten to catch it. It twists its neck and gives a weak bite. Beady eyes peer at me, its neck is near my finger and the tail droops from my hand. My prize goes into the bag.

Next day I find its name—"Mueller's Skink". It grows much longer, to forty centimetres, always has a small head and a black throat, but sometimes the white stripes are missing. It was so active that when I set it down to photograph it at home, it escaped. I now have a Mueller's Skink living under my house.

Pit-shelled Turtles (plate 75) live mainly in the freshwaters of the Fly and Strickland Rivers and adjacent swamplands in Papua, and in the southern rivers of Irian Jaya and rivers of Arnhem Land in Australia. Young turtles are black with a distinct ridge on the back and a saw-toothed carapace. They grow to become smooth-backed adults 60 cm in length. Irregular pitting occurs in the shell of adults. The tube nose may be an aid in breathing from beneath the surface or may prevent fruit pulp entering the nostrils while feeding. This turtle eats *Pandanus* fruit, mangrove seeds and assorted vegetation.

D'Albertis' Python (plate 76).

77

The largest snakes in Papua New Guinea belong to the python family. Heavy-bodied **Boelen's Pythons** (plate 80) from mid-mountain forests grow to six metres and are the heaviest of our snakes. However the Amethystine Python from lowland forests, reaching seven metres, is longer and slimmer. Both species live on the ground and include other snakes in their food, occasionally devouring snakes almost as long as themselves.

78

79

A small non-venomous snake from the lowlands, **D'Albertis' Python** (plates 76 and 78) rarely grows to more than three metres. Like most pythons it is stout and muscular. Two colour forms are shown: one from near Port Moresby is black above, white below; the other from near Lae is black only on the head and the remainder of the back is rich red-brown. The small pits in the lower jaw contain heat-sensing receptors which aid in locating warm-blooded prey at night.

Juvenile **Green Tree Pythons** may be either golden yellow with a darker pattern (plate 77 and back cover) or peppered brick-red with a lighter pattern of spots. At one metre, individuals change colour and adopt the bright green of the adult (plate 79).

80

81

82

Most boas are small species never more than a metre in length. Both New Guinea species, the **Ground Boa** (plates 81 and 82) and **Tree Boa** (plate 83) have a broad flattened head with prominent eye ridges. Unlike the **Death Adder** (plate 88) which is similar in shape to the Ground Boa, the body scales of the boas are strongly keeled and small scales cover their head. The group of Pacific boas includes five species ranging from eastern Indonesian islands along the north coast of New Guinea to the south-west Pacific region. They are harmless snakes varying in ground-colour from yellow-brown to dark brown with a chain of spots frequently forming a zig-zag along the back.

Three venomous snakes from different habitats are the **Brown Tree Snake** (plate 84), **Javan File Snake** (plate 85) and **Banded Sea Snake** (plate 86).

83

84

85

The Brown Tree Snake is a slender arboreal snake which may be two metres long yet hardly thicker than a thumb. Active at night and usually quick to strike, this species is nevertheless harmless to humans as its poison fangs are sited at the back of the mouth and its bite is inefficient.

The Javan File Snake is also a rear-fanged snake. Living in still or slow-moving fresh water this unusual species seeks its prey, mainly fish, in the water weeds. On land it is an unattractive flabby, beady-eyed snake which is extremely docile and does not attempt to bite.

Also docile is the Banded Sea Snake. A marine species, this is one of the few sea snakes which habitually come on to land. Small coral and limestone islands are favoured resting places during the day and clusters of up to twenty or more snakes may occasionally be found sleeping in a shady rock overhang.

86

87

Most dangerous of the New Guinea snakes are the fast moving, venomous front-fanged species. Little is known of the effects of bites of the endemic species and most deaths recorded are from species which evolved in Australia, and have populated southern and south-eastern Papua.

88

89

The **Papuan Taipan** (plates 87 and 90) is a particularly dangerous species which grows to over two metres. Dark above and pale below, with an orange or reddish stripe running along the back, the Taipan rarely occurs higher than 800 metres. Before shedding the skin the scale over the eye lifts and the hollow fills with a milky fluid, as pictured. At this time individuals are nervous and strike easily.

90

Above 800 metres only two dangerous snakes are likely to be found, the **Death Adder** (plate 88) and the Small-eyed Snake. These range almost as high as 2 000 metres. Both are widespread and much feared. The Death Adder is distinguished from the similar Ground Boa by its blunt snout, checkered pattern on the face and spine on the tail tip.

Similar in appearance to the dangerous front-fanged snakes are two rear-fanged tree snakes. The **Green Tree Snake** (plate 91) varies in colour from greenish to bluish, usually with yellow on the under surface. It is a relatively stout snake, growing to 150 cm. Related to this, the **Slender Tree Snake** (plate 92) is a pencil-thin species to 100 cm. Beautifully marked with white lips, a creamy white pattern along the back and pale blue between the scales, this is one of the most attractive snakes of Papua New Guinea. Both Tree Snakes are regarded as harmless.

Large-eyed and fast, the **Papuan Whip Snake** (plates 89 and 93) is a slender ground-dwelling species characteristic of the drier Papuan savannahs. Its bite is probably not fatal. It is best identified by the orange-brown tail.

91

92

93

94

95

Both species of crocodile from New Guinea occur in fresh water. The **Freshwater Crocodile** (plates 94 and 95) is restricted to rivers, swamps and lakes while the Saltwater Crocodile also lives in the ocean and saline estuaries. Hunting has depleted the total number of crocodiles in Papua New Guinea and the prized Saltwater Crocodile is not frequently seen.

Freshwater Crocodiles hatch at 30 cm and reach a maximum length of four metres. When young they feed mainly upon insects but gradually change their diet to include reptiles, mammals and birds as they mature. An aged Saltwater Crocodile may reach ten metres. Animals of this size are rare and the largest encountered today are usually no longer than six metres.

Green Anglehead (plate 96)

Angleheads are a distinctive group of dragon lizards which spend most of their lives in the trees. They are shy and not frequently seen as they stay motionless in the canopy or move around the tree-trunk to hide from an intruder.

97

98

The **Green Anglehead** (plate 96) grows to
30 cm. It can change its colour from bright
green with bluish on the face, to a dull brown
within a few minutes. It has a small throat
pouch and is rather like the Anole lizards of
Central America.

Largest of the ten species from New Guinea is
Godeffroy's Anglehead which grows to 120 cm.
It has flattened spines along the back similar
to the **Keeled Anglehead** (plates 97 and 99).
This species has a blotchy appearance due to
large scales on the side of the body contrasting
with the even, smaller scales. The **White-
cheeked Anglehead** (plate 98) lacks these
larger scales and possesses a distinctive check-
erboard pattern on the throat. Both these
Angleheads threaten an intruder by extending
the remarkable throat pouch and opening the
mouth to display a bright yellow interior and
yellow tongue.

99

101

100 Gekkos are soft skinned nocturnal lizards found in a variety of situations. The **Striped Gekko** (plate 100), **Loria's Gekko** (plate 101) and **Oceanic Gekko** (plate 104) live in lowland and mid-mountain rainforest and either actively seek out their insect prey on the trunks and branches of trees, or wait quietly for insects to come within reach. The Striped Gekko is the most handsome species, ranging from Indonesia east to the Solomon Islands. Loria's Gekko, here perched on a cut ginger stem, grows to 30 cm and is the longest New Guinea gekko. The Oceanic Gekko has delicate skin which peels off unless handled with care. Having no eyelids it must clean and lubricate its eyes with its tongue.

102 The **Emerald Monitor** (plate 102) and **Spotted Monitor** (plate 103) are two of the seven New Guinea monitors. All are fast-moving, and generally ground-dwellers. The slim Emerald Monitor, however, spends most of its time in trees. It occurs throughout lowland rainforests. Spotted Monitors prefer the vicinity of water in lowland forests and mangroves. The small-spotted pattern separates them from other New Guinea monitors.

103

104

105

Over one hundred species of skinks, ranging in size from the small and widespread *Carlia* species (15 cm) to the giant Prehensile-tailed Skink (100 cm) of the Solomon Islands, form the main family of lizards in New Guinea. **Ground Skinks** (plates 108 and 109) scurry through the litter in the rainforest seeking insect prey. They lay a small number of tiny white eggs in a hollow in the ground. The **Green Tree Skink** (plate 107) lives on the trunks and branches of trees, especially coconuts, along the north coast and on New Britain. Two individuals meeting go through an amusing head-bobbing, tail-wagging dance of recognition.

106

Mueller's Skink (plate 105) is a peculiar heavy-bodied, brightly-patterned ground lizard with a tiny pink-tipped nose. With its short legs and unusual wobbling gait it is most probably adapted to worm its way beneath the leaf litter. The unidentified skink (plate 106) with the beautiful rainbow sheen is also a leaf-litter species from the humid lowland forests of the Northern District.

107

108

109

FROGS

Though a diverse fauna, occupying a wide variety of habitats and many unique life-styles, the 160 species of frogs in Papua New Guinea are dominated by two families. The tree frogs, *Hylidae* and *Microhylid*, have roughly sixty and eighty species each. Tree frogs characteristically are long-legged species with webbed toes ending in a large disc. The two genera of this family can be separated by the shape of the pupil, which in *Litoria* is horizontal and *Nyctimystes* vertical.

The next most important family, Ranidae, similarly contains only two genera. *Rana* are frogs of the forest floor, stream beds and open areas adjacent to water in the savannah. The various species generally have a conspicuous fold down each side and frequently have bands on their hind legs. These frogs lay their eggs in water and the tadpoles hatch and metamorphose in the classical sequence. In contrast the ranids of the genus *Platymantis* are often boldly patterned with short ridges upon the back. The common species, *P. papuensis* has a number of different forms which all utter the same call. *Platymantis* lay their eggs in damp litter on the forest floor and tiny frogs hatch direct from the egg.

The remaining two families of frogs include the burrowing frogs of the family Leptodactylidae, which are mostly Australian forms found only in southern Papua. These have a well developed horny 'spade' on the hind foot and expertly burrow beneath the surface to avoid dry spells. Finally an introduced species, common in the lowlands, belongs in the family Bufonidae. This is the Marine Toad, *Bufo marinus*. Introduced originally to the Gazelle Peninsula prior to World War Two, to control the Sweet Potato Moth, it is now slowly colonising open areas in the lowlands of all Papua New Guinea.

FROGS—160 species, 5 families

Come and stand with me, thigh-deep in a swamp, and listen to the chorus of frogs. It is night, it has been raining and the frogs and mosquitoes are out. Here is the largest tree frog in the world, the White-lipped Tree Frog, sitting on a palm leaf calling its scratchety-ratchety call. Sitting on its head is a mosquito, no—two. Scratchety-ratchety he calls, bothered only when a mosquito lands on his eyelid and bores for his blood. Tree frogs are calling in a swamp, scratchety-ratchety, throat swelling, subsiding.

I remember a contrast in the village of Garu far from here. There were scratchety-ratchety calls but no tree frog. This was a Rana, *allied to the frogs of the Old World, brown, sharp-nosed, sitting upright. I heard the loud call from afar and tracked it down. It has heavy thighs; perhaps you can eat them. Then from the bananas I hear another call, strong, sharp, short. It is a small frog. But where is it? The call comes from here where my light shines, but where is the frog? Again comes the call; again there is no frog. It should be on the broad blade of the banana leaf. But where is it? Stand taller, track it down. Listen from this side, listen from that. It must be there. The sound keeps coming from the same place in the middle of the leaf. Puzzled I step back and the riddle is solved. It was there all the time, but not where its voice came from. It was on the banana leaf—not on the blade, but in the stem. The yellow stem was curled to a tube and inside was the frog, invisible. Its voice goes along the tube to emerge from the blade. A quick grasp with my hand and I catch the frog. It is small, one centimetre long, and mid-brown in colour. An* Oreophryne *from New Britain.*

But listen, scratchety-ratchety is calling again. Listen . . . scratchety-ratchety, scratchety-ratchety . . . fweep. Again . . . scratchety-ratchety, scratchety-ratchety . . . fweep. A whistling frog? Listen once more. Scratchety-ratchety, scratchety-ratchety . . . fweep. Next time I visit Garu I will bring my tape recorder so that people will believe this.

110

From the southern slopes of the Owen Stanley Range, at 1 300 metres, sixty kilometres north-east of Port Moresby, the **Sharp-snouted Tree Frog** (plate 110) is separated from all other tree-frogs in New Guinea by the long spine on the nose. It is four centimetres in length and well camouflaged with blotches and smudges of grey and brown.

The **White-lipped Tree Frog** (plate 111).

The largest tree frog in the world, the female **White-lipped Tree Frog** (plate 111) may reach over thirteen centimetres in body length. It is a widespread species in the savannah woodlands and forests of the lowlands to about 1 100 metres. Here it prefers to live near water, and voices its deep scratchety-ratchety call from a perch up to two metres above the ground. The frog in the picture is sitting on a cluster of banana flowers.

The tree frogs and ground frogs pictured show the variety of shape and colour in New Guinea frogs. The tree frog (plate 112) illustrates the enlarged toe disc typical of the genus *Litoria*. The completely webbed toes of this species are unusual. **The Gold-striped Tree Frog,** *Litoria vocivincens* (plate 114) and *Litoria bicolor* (plate 115) also show these discs. The small frog (plate 113) is typical of the ground-living genus *Cophixalus*. Smallest frogs in this genus are only fifteen millimetres in length. Related to *Litoria*, *Nyctimystes foricula* (plate 116) has less-developed discs on the toes and, in the bright light, the pupil is vertical like that of a cat. *Platymantis* frogs may come in a number of colour varieties. *P. papuensis* (plate 117) shown here may also be clearly striped on the back, unlike the blotchy pattern shown. *Phrynomantis lateralis* (plate 118) is a solitary inhabitant of rainforest litter related to *Cophixalus*. The three pictures on the left show frogs in typical calling positions.

113

114

115

116

117

118

119

Marine Toads (plate 119) were introduced
to the Gazelle Peninsula in 1936 to control
an outbreak of sweet potato moths which
were damaging crops. They were later
introduced to the mainland both north and
south of the mountain chain and are now
well-established throughout the lowlands.
A ground-living species, it prefers disturbed
grassy areas and frequents towns and
villages. Sparse large individuals occur also
in the rainforest. Scavengers, with a wide
range of animal food, the toads in the picture
are eating a dead Papuan Whip Snake.

A small short-legged tree frog of the lowlands of Papua *Litoria congenita* (plate 120)
is shown here asleep two metres above the
ground.

120

Long slender toes and a prominent ridge from the eye and down the side distinguish frogs of the genus *Rana*. *Rana papua* (plate 121) is rarely found away from the ground. It is a lowland forest species which seeks insect food in the leaf litter. Most *Rana* have barred legs, as shown. This is the same genus as the common ground frogs of the northern hemisphere.

Sleeping in a similar situation to the tree frog opposite, the small frog with a striking head pattern (plate 122) is probably a juvenile **Ranid** of the genus *Platymantis*. These frogs on New Britain tend to climb higher than on mainland New Guinea where it is rare to see them above ground level.

123

124

125

126

127

The **Black-spotted Tree Frog**, *Litoria iris*, (plate 123) occurs from 1 500 to 2 500 metres. The **Slender Tree Frog**, *L. impura*, (plate 124) is a lowland forest species, as is the **Gold-striped Tree Frog** (plate 127) *L. vocivincens*. The **Green-crowned Tree Frogs** of the *L. "arfakiana"* group (plate 125) include a number of species which live near mountain streams to 2 000 metres. They are characterized by variable amounts of green on the crown.

Rana grisea (plate 126) is a strongly marked ground frog reaching a higher altitude than the Papuan Rana shown earlier.

INSECTS AND SPIDERS

The insect life of Papua New Guinea is rich and varied. It is not possible to estimate the number of species, but beetles number over 30 000 species and moths and butterflies over 6 000 species. In addition, flies, ants and wasps form other abundant groups. In all, it is probable that between 50 000 and 100 000 insect species occur in this country.

The most conspicuous insects are the ever-present variety of butterflies and dragon-flies. Papua New Guinea is the home of the world's largest butterfly. The female Alexandra Bird-wing, found in a limited area near Popondetta in northern Papua, spans over twenty-five centi-metres. Here too is one of the world's largest moths—the Hercules Moth. These are not rare insects, for in their preferred habitat they may readily be seen by even the casual observer.

Insects are most prolific in the lowlands and misty mountain rainforests, particularly in swampy areas. Here, beneath the canopy, an array of insects will be found on the fig trees, *Eugenia* flowers, nutmegs, palms and rattans. During the day, beetles and butterflies seek food beneath the canopy. At night a host of hawk moths and fruit-sucking moths appear. In the clearings in the forest, especially in the *kunai* and *Saccharum* grasslands, butterflies and grass-hoppers familiar in northern Australia abound.

The spiders of Papua New Guinea are varied too. There are well over five hundred species dominated by the familiar Orb Weavers—which include many colourful species building their webs in a variety of shapes and sizes—and the Jumping Spiders. The latter also can be colour-ful. The large *Mopsus mormon* is a brilliant leaf green with a pattern of false eyespots on the back of the female, or black face with a bushy moustache in the male. Iridescent blues and greens and a rusty red are also common colours among the Jumping Spiders.

I first saw the male beating the female down with his wings. He was a plain brown butterfly which flew above and behind the sparsely spotted female and forced her to settle. Then, hovering above her to keep her in place, he quickly darted at her and flicked his abdomen forward to fertilize her. These were Crow Butterflies Euploea, and it was in the savannah near the Bensbach River in Papua that I saw their courtship flight. Later, in other parts, I was to witness it again, with other species—the Orchard Butterfly, the dark male fluttering with swift wingbeats above a pallid female, coercing her to settle; and with the Birdwings, an emerald green male partnering his darker mate in a spiralling dance of fancy. Photographing them was difficult.

The dragon-flies were easier though. On the edge of the forest, red-winged and black-winged species settled on the kunai blades. Shy, restless, always ready to take flight when approached, they seem difficult to photograph, but not if you know the secret. They can be plucked from their perch at night when they sleep, or in the early morning when the grass is still wet with the cold and the dew sparkles on their wings.

Dragon-flies, sometimes in vast swarms, with hundreds of shimmering wings quivering in the sun, dart here and there, perhaps in a courtship dance, perhaps after an abundance of food. Dragon-flies are found with red wings, black wings, blue bodies, crimson bodies, checkered wings, auburn eyes, banded eyes, in a multitude of shapes and sizes and colours.

INSECTS—50-100 000 species, numerous families
SPIDERS—500 species, 30 families

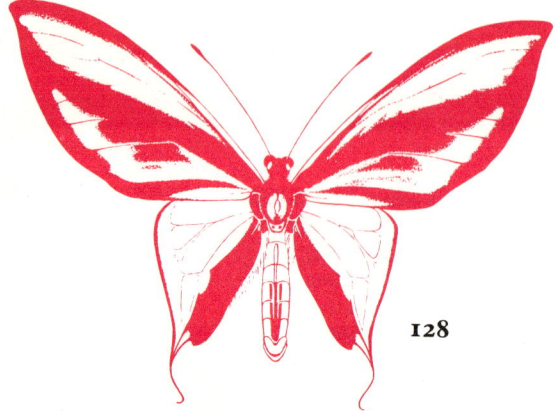

128

The male **Paradise Birdwing** (plate 128) has delicate tails on the hind wing. Female birdwings are duller—black, cream, yellow and some red. The largest butterfly in the world, the female Alexandra's Birdwing spans twenty-seven centimetres across the wings.

The **Common Birdwing** (plate 129).

130

131

Fully-grown birdwing caterpillars are armed with soft spines which offer little protection. Shown here slightly larger than life, the caterpillar of a **Tailed Birdwing** (plate 130) feeds on an *Aristolochia* climber. The pupa of the **Common Birdwing** (plate 131), life-size, is slung beneath a leaf or convenient twig. A female **Common Birdwing** (plate 132) contrasts with the larger but duller **Victoria Birdwing** female (plate 133) from Bougainville.

132

133

134

135

Most spectacular of all Australasian butterflies, the Birdwings occur in lowland and mid-mountain forests in Papua New Guinea. Males are brilliant green, yellow and black, as pictured in the **Common Birdwing** (plate 129).

In the same family as the birdwings the **Big Greasy** (plate 134) is found from northern and eastern Australia to the open grassy savannahs of Papua. The male, pictured above feeding upon a wild *Stachytarpheta* flower, is larger and brighter than the dull, brownish female. The female is unusual in the swallowtail family as it has transparent wings.

Also found in Australia but restricted there to coastal areas of north-eastern Queensland, the **Grass Yellow Butterfly** (plate 137) is found throughout New Guinea from the western Papuan islands to the Solomons. But it is not common. Though the male is always golden yellow, the female varies from white through yellow to smudgy grey on the upper surface, always with broad black margins to the wings.

Prothoe australis (plate 135) flies beneath the canopy of the rainforest and habitually rests head down upon a tree trunk, as shown. The caterpillar feeds upon wild citrus trees, *Evodia* species, of the under-storey. This species is widespread on the mainland and eastern islands across to New Ireland.

The **Owl Butterfly** (plate 136) family occurs mainly in the Indo-Australian region but barely enters Australia. Erratic flying individuals are characteristic of lowland and mid-mountain rainforest, always in dimly-lit areas and most frequent in swamp forest. The caterpillars are hairy and gregarious, both unusual characteristics for most species of butterflies.

136

137

Other rainforest butterflies, *Symbrenthia* (plate 138), and the **Five-bar Swordtail** (plate 139) prefer open sunny clearings while the **Hamadryad** (plate 141) is a slow flier beneath the canopy. *Symbrenthia* is a solitary, restless and fast-flying species from mid-mountain forests. The Swordtail in contrast is a social species with scores of individuals clustering upon a suitable food source, or attracted to the bright patterns of dead males lying on the ground.

139

138

140

Though it flies alone during the day, the Hamadryad congregates in loose groups to sleep. Eight butterflies were sleeping within 15 cm of each other in the swamp forest where the photograph was taken near midnight.

The **Crow Butterfly** (plate 140) is one of a confusing group of similar species common throughout the lowlands. The male is pictured in courtship flight. It flies above and 10 cm behind the female, beating her to the ground before mating.

141

142

143

144

Day-flying moths, brightly coloured like butterflies, may easily be mistaken if the simple antennae are not seen. Butterflies have a clubbed tip to each antenna. The two small moths of the family *Calledulidae* (plates 142 and 143) are lowland forest species. The species on the right is widespread on the mainland and though always orange and navy-blue the pattern varies throughout its range.

The large slow-flying **Uranus Moth** (plate 144) resembles a swallow-tail butterfly with its tailed hindwings. It flies in open areas, in lowland savannahs, rainforest and mid-mountain forest. At certain times of the year it is particularly common, even flying into houses in the larger towns. It has a wingspan of 15 cm and flies during both night and day.

145

146

A large and striking group of **Fruit-sucking Moths** (all photos this page) occur throughout New Guinea. They are attracted to rainforest fruit and their stout mouthparts can penetrate the thickened skin of fruit such as the wild figs (plates 146 and 148) to suck the sweet juices inside. As the *Ficus* flowers are hidden internally within the "fruit" capsule these moths probably play an important part in cross-pollination. Fruit-sucking moths are generally fast-flying species with delta wings. Many rainforest species have intricately patterned fore-wings of greens, browns and blacks which cover a brightly contrasting hindwing. The caterpillar typically has eyespots as shown (plate 147).

The bright scavenging **Orange-barred Moths** (plate 149) are shown feeding upon body juices released from a half-buried decaying Marine Toad. This moth is common in mid-mountain areas.

147

148

The day-flying moth, *Alcides zodiaca* (plate 150), belongs to the same family as the Uranus Moth (plate 144). It is a remarkable mimic of the swallowtail butterfly, *Papilio laglaizei*, differing only slightly in distribution and intensity of colouration on the upper surface. The undersurface, which is not usually exposed, varies more.

Moth and butterfly caterpillars vary tremendously in size and shape. The four shown here have been selected to illustrate variety in adornment, from the naked skin of the yellow **Geometer caterpillar** (plate 152) to the thick hairs of the large species (plate 153) and the clustered hairs of the **Toothbrush caterpillar** (plate 154).

155

156

157

158

Tree-Ants and their bulky leaf nests are a characteristic sight in lowland forests, particularly savannah woodland.

Always busy, constructing a new nest chamber (plate 155), gathering food for the colony (plate 156) or tending their farm animals (plates 157 and 158) these ants are the epitome of social development.

Adults produce no silk. While some workers hold leaf edges together, others carry larvae from the nursery to spin the silk which binds the leaves of the nest.

Both the caterpillar of the **Blue Butterfly** (plate 157) and the **Scale Insects** (plate 158) exude a sweet fluid from pores in their back which the ants consume in return for protection.

159

160

161

162

163

164

Dragonflies (left column) and **damselflies** (right column) always fly relatively close to water. Coupled animals, such as those in a mating flight (plate 160), dash and dart about water and either perch so that the female may lay eggs in water weeds, or dip into the surface to lay eggs.

Dragonflies are typically heavy-bodied fast-flying species which always rest with their wings held flat at right angles to the body.

Damselflies are delicate and slow-flying. They hold their wings tented above the body at rest.

165

166

167

168

Insect life in the forests of New Guinea is rich and varied. **Painted Weevils** (plate 165) form a distinctive group of colourful beetles from sea-level into the upper areas. There are some 5 000 species of weevil in New Guinea. **Flower Beetles** (plate 166) are active day-flying species which noisily buzz about flowering trees and fruit such as *Pandanus* in search of sweet food. Their bright metallic colours are prized by Highlanders who make colourful headbands from the shells.

The small number of Jewel Beetles in New Guinea reflect the brilliant colouration the family is known for in Australia. **D'Albertis Jewel Beetle** (plate 167) is a bright burnished species with false eyespots on the thorax.

Like the Hamadryad butterflies pictured earlier the small **Solitary Bees** (plate 168) cluster into groups to sleep at night. About fifty individuals were discovered asleep on this lowland rainforest climber, each clinging tightly with its jaws clamped on the plant to ensure a sound sleep.

169

170

A selection of spiders from four families; the **Orb Weaver** (plate 169)—family Argiopidae —spins a traditional wheel-like flat web and waits for prey to blunder in. In this case a small lizard, probably *Carlia* or *Lygosoma* species became ensnared, was rapidly bound up into a silken girdle, then consumed at leisure as the orb-weaver pumped digestive juices into its body before sucking in the nutrients from the broken-down tissues.

The Lichen Spider (plate 170)—family *Sparassidae*—in contrast, is a hunter on tree trunks and boughs in the forest. It is beautiful-ly camouflaged to resemble lichen and is difficult to detect unless it moves. Many spiders of this family bear different patterns and textures to hide them on the bark of the trees they prefer.

171

172

The **Bird-eating Spider**, shown here life-size (plate 171)—family *Theraphosidae*—is fearsome of appearance and is dangerous. It lives in an open-topped hole in the ground. Small trip lines radiating from the hole alert the hidden spider to the presence of prey. Though usually shy and not often seen, occasional individuals leave the shelter of their burrow and wander widely. The bite is extremely painful and could probably kill a child.

Most delightful of the spiders are the alert, often comical **Jumping Spiders** (plate 172)—family *Salticidae*. *Mopsus* males have black forelegs and a fine white brush of hairs behind the eyes, unlike the female which is all green with a pattern on the back containing false eyespots. Jumping spiders search out and stalk their prey before suddenly jumping to capture it.

PLANTS

The main group of plants in Papua New Guinea is the orchid family. Almost three thousand species occur here, over 130 genera dominated by the epiphytic *Dendrobium* and *Bulbophyllum*.

In the lowlands, in the gallery forest lining meandering rivers and ephemeral creeks, in gaps in the canopy of the rainforest, on the uppermost branches of the tallest trees, wherever there is light there are orchids. Here the "antelope-group" of *Dendrobium* is common, with its thin sepals raised above the lip like the horns of an antelope. Higher in the mountains a second group of *Dendrobium* replaces the antelope orchids; this group is characterized by a triangular-shaped flower with a prominent central lip. These *Oxyglossum* orchids prefer the damper and cooler air of the mountains.

The magnificent splendour of the orchids is displayed in a setting of unbelievable complexity. The tropical lowland rainforest, as an ecosystem, is one of the most complex of habitats to be encountered upon the earth. Rarely do pure stands of rainforest trees occur as in the temperate zones. One hectare of forest may contain over five hundred species of plants, all competing for the sun; yet within a kilometre a separate hectare may contain another five hundred species and little duplication. This pattern is repeated again and again.

Higher, in the cooler mid-mountain altitudes from 500 to 2 500 metres, the forests may be dominated by oak trees, or in the belt 800 to 2 000 metres by pines, the Klinkii Pine and Hoop Pine being best known. Reaching higher still, to 3 000 metres, stands of Antarctic Beech occur. But although in one area the forests may occur in relatively pure stands this is not always the case; the rule that applies in one area does not apply in another and the broad generalisation of altitudinal distribution of forest trees must be recognised and modified as necessary.

PLANTS—10 000 species, 350 families

173

From a distance I could not identify the orange cluster on the side of the tree fern. It was so bright. Perhaps it was some fruit of a vine. Clamber twenty metres up the slope and pause for breath. Here at 3 000 metres on Mt. Victoria the air is thin. A further ten metres and the cluster is just above me. Bright bell flowers droop from shining green leaves. It is an orchid. Later, in the comfort of Port Moresby, I find out its name—Dendrobium flammula.

Lower down below Murray Pass, between Mt. Victoria and Mt. Albert Edward, I see another orchid. This time it is on the ground, growing from an earthen bank and looking like a cluster of cherries. This is Epiblastis basalis. *I will always think of it as the "Cherry Orchid".*

Many of the orchids in Papua New Guinea live high on a tree. They press their roots close to the bark and shower sprays of flowers below them. Dendrobium discolor is widespread and in many varieties. The largest spray of any orchid I have seen came from such a plant on the banks of the Bensbach River. Each stem ended in a spike—the largest being almost two metres long with over a thousand flowers and the shortest about fifty centimetres. In all, it was quite an armful.

The two thousand, seven hundred species of orchid on this island are almost a fifth of its plants. The smallest flower is minute, barely as large as this O, the largest is the size of my hand. The tree dwellers are Dendrobium, Bulbophyllum *and* Oberonia; *the ground dwellers are* Spathoglottis, Calanthe *and* Nervilia. *Search well in the grass, for the kunai is thick and you may miss the ground dweller. Search well in the forest for the canopy is high, but in a break where the sunlight shines, there you may see the tree dweller.*

The Rhododendrons, *too, like the light. In the open canopy over the forest you may see them clinging to tree trunks and rivalling the orchids in their search for the sun. They have thinner stems than the orchids, round leaves and red, yellow and green clustered bell flowers. The high mountains have many. As with the birds, the animals, the orchids, many mountains have their own species. Here, cut off from the world by deep valleys, have evolved the multitudinous varieties we see today.*

Rhododendron aurigeranum (plate 173) is a mountain species found above 1200 metres in the highlands. Its dull gold flowers, with a delicate pink flush towards the tip of each petal, reflect the colours characteristic of the genus: shades of orange, pink, purple, red and yellow.

The genus *Bulbophyllum* (plate 174).

175

176

177

The orchid pictured in plate 174, belongs to the genus *Bulbophyllum*. This is the largest of all orchid genera, with about 2000 species worldwide and 600 in New Guinea.

Although varied in shape, as shown here, and in size from minute flowers 5 mm across to the large-flowered species shown natural size (plate 176), *Bulbophyllum* orchids seem to have an indefinable similarity which permits the beginner to recognise them. Many have shades of purple, often with long drooping sepals (plate 175), or symmetrical in shape such as the **Star Orchid** (plate 177) or the species on the previous page.

Bulbophyllum orchids are mainly tropical, from Africa through tropical Asia and New Guinea into the south-west Pacific. The genus is poorly represented in South America and there are twenty-one species in northern and eastern tropical areas of Australia.

178

179

180

The orchids on this page are from the two extremes of forest habitat in Papua New Guinea. The **Cherry Orchid** (plate 178) occurs in stunted moss forest above 3 000 metres, an area of prevailing cold and misty climate. Early morning clouds disperse for perhaps two hours, giving the forest limited sunshine before the mid-morning and afternoon cloudbanks build up and clothe the forest in a blanket of grey.

In contrast the **Ground Orchid** (plate 179) and **Pink Ground Orchid** (plate 180) occur in the steamy lowlands and in forests with high temperature and high humidity. Though the Ground Orchid blossoms beneath the canopy, and the Pink Orchid in open disturbed habitat, both extend to about 1 200 metres above sea-level.

Dendrobium orchids, with over 1500 species in the world are restricted to tropical Asia-east to Fiji and south to Australia: New Guinea claims over 300 species rivalling *Bulbophyllum* in importance, and these show equal diversity in shape and colour.

Though chiefly a lowland forest genus there are certain groups of related species which prefer the higher mountains. The **Flame Orchid** (plate 181) illustrates the flat, hand-like appearance of many of these mountain dwellers. Brilliant orange, advertising itself strongly on a tree-fern in the alpine grassland above the moss forest, this plant was photographed at 3500 metres in Murray Pass between Mt. Scratchley and Mt. Albert Edward.

Most species of orchids attract the insects, and possibly birds which pollinate them, by their distinctive colours. Few are scented. One such, the **Scented Orchid** (plate 182) has an odour reminiscent of raspberry jam. The large flowers, shown almost life size, arise in loose clusters from long drooping stems. The common colour variety is shown: a delicate mauve form is also known.

The **Bottlebrush Orchid** (plate 183) is widespread throughout New Guinea and occurs in Australia in the tropical forests of Queensland. Its colour varies from pale green through white to light or dark pink. Individual flowers deepen in colour as they mature. The flowers are tightly bunched near the end of a stem, giving the species its common name.

Shy Orchids (plate 184) are restricted in New Guinea to a small area north and west of Port Moresby. They prefer the seasonally dry climate of this area with little or no rain during the cooler winter months. Growing on trees, the Shy Orchid exposes itself to the sun and the large flowers nod in small clusters at the end of a long stem. Some plants have dwarf flowers, each 4 cm across, others may reach 10 cm. The overall pattern of colour is similar to that of the Bottlebrush Orchid above.

182

183

184

185

Prettiest, and one of the most spectacular flowers of New Guinea, *Rhododendron zoelleri* (plate 185) grows at 1 000 metres and above. It is a widespread plant which grows mainly on trees but may be a free standing shrub. The sight of a shrub of this species, covered in saucer-sized blossoms, is unforgettable.

Macgregor's Rhododendron (plate 186) is also a common and wide ranging species, found mainly in mid-mountain forests and moss forest. Colour varies from the dull golden variety pictured through pink to a variety like a miniature of the *R. zoelleri*.

There are 157 species of *Rhododendron*, all but two found nowhere else but New Guinea. Largest of the genus is the giant *R. leucogigas* from Irian Jaya; smallest are alpine species from the mountain-top heathlands.

Rhododendrons are now widely cultivated, particularly in the temperate regions of our world, where the climate is similar to their mountain homes: this is a contribution from New Guinea to the gardeners of the world.

187

188

189

The buds of the *Eugenia* tree (plate 187) illustrate a phenomenon quite common in tropical forests but rarely seen in temperate regions: buds, flowers and fruit growing directly from the trunk of a tree instead of among the leaves in the canopy. This is especially true with many of the wild *Ficus* species, dense clusters of figs providing abundant food for fig parrots, fruit-sucking moths, wasp larvae etc.

Both red fruit pictured (plates 188 and 189) are from a moss forest at Tomba near Mt. Hagen. The delicate red berries are the fruit of a ground-cover plant which creeps moss-like over the soil in suitable damp locations. The orange jelly-like capsule of the second fruit splits to reveal the seeds inside. This is a plant of the shrub layer, growing to one metre in height beneath the conifer canopy.

Flame-of-the-Forest (plate 190) is an apt name for the bright flowers of a lowland rain-forest climber. From the air brilliant splashes of red advertise these plants, which climb hundreds of feet into the canopy with a ground stem as thick as a man's leg. Red, green, white and yellow flowering forms are known.

191

192

193

Poor sandy soils, lacking in nitrogen, frequently promote the development of insect trapping plants. These catch and digest insects, absorbing the nitrogen broken down from the animal protein to build their own tissues. Both species pictured are from swampy soils in the Western District. The **Pitcher Plant** (plates 191 and 193) has a small amount of water in the pitcher developed at the end of certain leaves. This contains an enzyme which slowly digests insects, mainly ants, which are attracted to sweet fluid near the lip and tumble into the pitcher.

The **Sundew** *Byblis* (plate 192) traps insects on the tacky globules at the end of hairs on the leaves. These then bend slowly and bring the body against special digestive hairs which absorb the nutrients.

INDEX—PLATES

This section lists the plates, with common name, scientific name and brief notes on each species. I prefer to use photographs only of free-living, wild animals but this is virtually impossible with the majority of Papua New Guinea animals. The point of capture of captive animals is given.

49. **Brown Owl** *Ninox theomacha*. Uncommon. Lowland savannah and forest. Baiyer River. Captive.

50. **Mid-mountain Eupetes** *Eupetes castanonotus*, male. Uncommon. Mid-mountain forest. Sogeri. Captive. *Photo : B. J. Coates.*

51. **Lowland Eupetes** *Eupetes caerulescens*, female. Common. Lowland forest. Brown River near Port Moresby. Captive. *Photo : B. J. Coates.*

52. **Ajax Quail Thrush** *Cinclosoma ajax*, male. Uncommon. Papuan lowland rainforest. Sogeri near Port Moresby. Captive. *Photo : B. J. Coates.*

53. **Lowland Mouse Babbler** *Crateroscelis murina*. Common. Lowland and mid-mountain forest. Sogeri. Captive. *Photo : B. J. Coates.*

54. **Blue Wren Warbler** *Todopsis cyanocephala*, male. Common. Lowland rainforest. Port Moresby. Captive. *Photo : B. J. Coates.*

55. **Lowlands Flatbill** *Machaerirhynchus flaviventer*. Common. Lowland rainforest. Brown River near Port Moresby. Captive. *Photo : B. J. Coates.*

56. **Sooty Thicket Fantail** *Rhipidura threnothorax*. Uncommon. Lowland swampforest. Brown River near Port Moresby. Captive. *Photo : B. J. Coates.*

57. **Rufous-backed Fantail** *Rhipidura rufidorsa*. Common. Lowland and mid-mountain forest. Mt Lawes near Port Moresby. Captive.

58. **Spot-winged Monarch** *Monarcha guttula*. Common. Lowland rainforest. Mt Lawes near Port Moresby. Captive.

59. **Olive Microeca** *Microeca flavovirescens*. Common. Lowland rainforest. Kuriva River near Port Moresby. Captive.

60. **Frill-necked Flycatcher** *Arses telescophthalmus*, male. Common. Lowland forest. Mt Lawes near Port Moresby. Captive.

61. **King Bird of Paradise** *Cicinnurus regius*, male. Common. Lowland forest. Mt Lawes near Port Moresby. Captive.

62. **Superb Bird of Paradise** *Lophorina superba*, female. Common. Mid-mountain forest. Wau. Captive.

63. **Brown Sicklebill Bird of Paradise** *Epimachus meyeri*, female. Uncommon. Mid-mountain forest. Baiyer River. Captive. *Photo : B. J. Coates.*

64, 65. **Raggiana Bird of Paradise** *Paradisaea raggiana*, male in display. Common. Lowland rainforest. Baiyer River. Captive.

66, 67. **Magnificent Riflebird** *Craspedophora magnifica*, 66—female, 67—male. Common. Lowland rainforest. Kuriva River near Port Moresby. Captive.

68. **White-eared Catbird** *Ailureodus buccoides*. Common. Lowland rainforest. Brown River near Port Moresby. Captive.

69. **Rufous Babbler** *Pomatostomus isodori*. Common. Lowland rainforest. Mt Lawes near Port Moresby. Captive.

70. **Long-billed Honeyeater** *Melilestes megarhynchus*. Common. Lowland rainforest. Mt Lawes near Port Moresby. Captive.

71. **Rusty Pitohui** *Pitohui ferrugineus*. Common. Lowland rainforest. Mt Lawes near Port Moresby. Captive.

72. **Brown Xanthotis** *Xanthotis chrysotis*. Common. Lowland savannah and rainforest. Mt Lawes near Port Moresby. Captive.

73. **Grey-headed Whistler** *Pachycephala griseiceps*. Uncommon. Lowland rainforest. Mt Lawes near Port Moresby. Captive.

74. **Straight-billed Honeyeater** *Timeliopsis griseigula*. Uncommon. Lowland rainforest. Kuriva River near Port Moresby. Captive.

75. **Pit-shelled Turtle** *Carettochelys insculpta*. Common but restricted distribution. Fly River. *Drawing : Vali Herzer.*

76. **D'Albertis' Python.** *Liasis albertisi*. Common. Lowland and mid-mountain rainforest. Brown River near Port Moresby. Captive.

77. **Green Tree Python** *Chondropython viridis*, juvenile. Common. Lowland rainforest. Brown River near Port Moresby. Captive.

78. **D'Albertis' Python,** see 76. Colour variety from Lae. Captive.

79. **Green Tree Python,** see 77. Adult. Brown River near Port Moresby. Captive.

80. **Boelen's Python** *Liasis boeleni*. Uncommon. Mid-mountain rainforest. Woitape. Captive.

81, 82. **Ground Boa** *Candoia asper*. Common. Lowland rainforest, mainly northern. Keravat, East New Britain. Captive.

83. **Tree Boa** *Candoia carinata*. Common. Lowland rainforest, mainly northern. Keravat, East New Britain. Captive.

84. **Brown Tree Snake** *Boiga irregularis*. Common. Lowland rainforest. Garu, West New Britain. Captive.

85. **Javan File Snake** *Acrochordus javanicus*. Common but restricted distribution in swamps, rivers and streams of south-western Papua. Weam. Captive.

86. **Banded Sea Snake** *Laticauda laticaudata*. Common. Shallow Papuan oceans. Bava Island near Port Moresby. Wild.

87. **Papuan Taipan** *Oxyuranus scutellatus*. Uncommon. Lowland rainforest in Papua. Brown River near Port Moresby. Captive.

88. **Death Adder** *Acanthophis antarcticus*. Uncommon. Lowland rainforest. Brown River near Port Moresby. Captive.

89. **Papuan Whip Snake** *Demansia olivacea*. Common. Papuan savannah. Port Moresby. Captive.

90. **Papuan Taipan,** see 87.

91. **Green Tree Snake** *Dendrelaphis punctulatus*. Common. Savannah and lowland rainforest edge. Port Moresby. Captive.

92. **Slender Tree Snake** *Dendrelaphis calligaster*. Uncommon. Savannah and lowland rainforest. Port Moresby. Captive.

93. **Papuan Whip Snake,** see 89.

94, 95. **Freshwater Crocodile,** *Crocodylus novaeguineae*. 94—Newly hatched Port Moresby. 95—Adult, 5 metres, Daru. Common. Lowland rivers, swamps, lakes. Captive.

96. **Green Anglehead** *Goniocephalus modestus*. Common. Widespread in lowland rainforest. Kimbe, West New Britain. Wild.

97. **Keeled Anglehead** *Goniocephalus dilophus*. Common. Lowland rainforest. Daru. Captive.

98. **White-cheeked Anglehead** *Goniocephalus papuanus*. Uncommon. Lowland rainforest. Long Island. Captive.

99. **Keeled Anglehead,** see 97.

100. **Striped Gekko** *Gekko vitattus*. Common. Widespread in lowland rainforest and villages. Kimbe, West New Britain. Captive.

101. **Loria's Gekko** *Cyrtodactylus loriae*. Common. Lowland and mid-mountain rainforest. Baiyer River. Wild.

102. **Emerald Monitor** *Varanus prasinus*. Common. Papuan savannah and lowland rainforest. Daru. Captive.

103. **Spotted Monitor** *Varanus indicus*. Common. Widespread in lowland rainforest and savannah. Port Moresby. Captive.

104. **Oceanic Gekko** *Gehyra oceanica*. Common. Lowland rainforest, villages, towns. Mt Lawes near Port Moresby. Wild.

105. **Mueller's Skink** *Lygosoma muelleri*. Uncommon. Lowland rainforest. Brown River near Port Moresby. Captive.

106. **family Scincidae.** Not identified. Lowland rainforest. Popondetta. Captive.

107. **Green Tree Skink** *Dasia smaragdina*. Common. North coast lowland rainforests, plantations, gardens. Popondetta. Captive.

108, 109. **Ground Skink** *Carlia* species. Common. Lowland savannah and rainforest. Brown River near Port Moresby. Wild.

110. **Sharp-snouted Tree Frog** *Litoria prora*. Uncommon, restricted distribution. Mid-mountain streams in Central District. *Drawing :* Vali Herzer.

111. **White-lipped Tree Frog** *Litoria infrafrenata*. Common. Lowland rainforest, villages, towns. Port Moresby. Wild.

112. **Tree Frog** *Litoria amboinensis*. Uncommon. Lowland swamp forest. Brown River near Port Moresby. Captive.

113. **Ground Frog** *Cophixalus cheesmani*. Uncommon. Widespread in lowland rainforest. Popondetta. Wild.

114. **Gold-striped Tree Frog** *Litoria vocivincens*. Common, restricted distribution in Central District. Brown River near Port Moresby. Wild.

115. **Tree Frog** *Litoria bicolor*. Uncommon, restricted distribution in south-western Papua. Weam. Wild.

116. **Yellow-spotted Tree Frog** *Nyctimystes foricula*. Common. Mid-mountain forest. Wild.

117. *Platymantis papuensis*. Common. Lowland rainforest. Garu, West New Britain. Wild.

118. *Phrynomantis lateralis*. Uncommon. Widespread in lowland and mid-mountain forest. Popondetta. Captive.

119. **Giant Toad** *Bufo marinus*. Introduced from Hawaii. Common. Widespread in lowlands particularly villages and towns. Port Moresby. Wild. *Photo :* B. J. Coates.

120. **Tree Frog** *Litoria congenita* asleep. Widespread lowland rainforest and savannah. Mt Lawes near Port Moresby. Wild.

121. **Papuan Rana** *Rana papua*. Common. Lowland savannah and rainforest. Morehead. Wild.

122. Not identified. Asleep. Lowland rainforest. Garu, West New Britain. Wild.

123. **Black-spotted Tree Frog** *Litoria iris*. Common. Mid-mountain rivers and ponds. Tari. Captive.

124. **Slender Tree Frog** *Litoria impura*. Uncommon. Lowland swamp forest. Mt Lawes near Port Moresby. Wild.

125. **Green-crowned Tree Frog** *Litoria "arfakiana"*. Common. Mid-mountain streams and ponds. Captive.

126. **Mountain Rana** *Rana grisea*. Common. Midmountain rivers and streams. Wau. Wild.

127. **Gold-striped Tree Frog,** see 114.

128. **Paradise Birdwing** *Ornithoptera paradisea*. Common. Northern mid-mountain forests. *Drawing :* Vali Herzer.

129. **Common Birdwing** *Ornithoptera priamus*, male. Common. Lowland rainforest. Brown River near Port Moresby. Captive.

130. **Birdwing** *Ornithoptera meridionalis* caterpillar. Uncommon. Southern lowland rainforest. Brown River near Port Moresby. Wild.

131. **Common Birdwing,** see 129. Pupa. Captive.

132. **Common Birdwing,** see 129. Female. Baiyer River. Wild.

133. **Victoria Birdwing** *Ornithoptera victoriae*, female. Common but restricted distribution. Lowland rainforest. Bougainville. Captive.

134. **Big Greasy** *Cressida cressida*, male. Common. Papuan savannah. Mt Lawes near Port Moresby. Wild.

135. *Prothoe australis*. Uncommon. Lowland rainforest. Brown River near Port Moresby. Wild.

136. **Owl Butterfly** *Taenaris catops*. Common. Lowland rainforest. Brown River near Port Moresby. Wild.

137. **Grass Yellow Butterfly** *Eurema candida*. Common. Lowland grassland. Port Moresby. Wild.

138. *Symbrenthia* species. Uncommon. Mid-mountain forest. Baiyer River. Wild.

139. **Five-bar Swordtail** *Graphium aristeus*. Common. Open areas in Papuan savannah. Brown River near Port Moresby. Wild.

140. **Crow Butterfly** *Euploea* species in flight. Common. Papuan savannah and grassland. Weam. Wild.

141. **Hamadryad Butterfly** *Tellervo zoilus* asleep. Common. Lowland rainforest. Brown River near Port Moresby. Wild.

142, 143. **Day flying moths.** Not identified. Common. Lowland rainforest. 142—Garu, West New Britain. 143—Brown River near Port Moresby. Wild.

144. **Uranus moth** *Nyctalemon* species. Common. Open areas in lowland and mid-mountain rainforest. Wau. Wild.

145, 146. **Fruit-sucking moths** family Noctuidae. Uncommon. Lowland rainforest. Mt Lawes near Port Moresby. Wild.

147. **Caterpillar of Fruit-sucking moth** *Othreis* species. Common. Lowland rainforest. Popondetta. Captive.

148. **Fruit-sucking moth,** as for 145.

149. **Orange-barred Moth** *Milionia isodoxa*. Widespread in mid-mountain clearings. Wau. Wild.

150. **Day flying Moth** *Alcides zodiaca*. Widespread in lowland and mid-mountain forest clearings. Baiyer River. Captive.

151-154. **Moth caterpillars.** 151, 154—Port Moresby. 152, 153—Popondetta. All wild.

155-158. **Tree Ants** *Oecophylla smaragdina*. Widespread in lowland forest and savannah. 155—Brown River near Port Moresby. 156—Weam. 157—with Lycaenid butterfly *Narathura* species. Port Moresby. 158—Mt Lawes near Port Moresby. All Wild.

159. **Barred Dragonfly** *Orthetrum sabinum*. Common. Lowland savannah. Port Moresby. Wild.

160. **Damselfly.** Not identified. Mating flight. Common. Near water in lowland savannah and rainforest. Brown River near Port Moresby. Wild.

161. **Red-winged Dragonfly** *Neurothema stigmatizans*. Common. Lowland savannah and rainforest clearings. Port Moresby. Wild.

162. **Black-winged Damselfly** family Chlorocyphidae. Common. Lowland savannah and rainforest clearings. Port Moresby. Wild.

163. **Dragonfly** *Rhyothemis phyllis*. Uncommon. Lowland savannah. Port Moresby. Wild.

164. **Damselfly.** Not identified. Common. Near water in lowland savannah and rainforest. Port Moresby. Wild.

165. **Painted Weevil** *Eupholus* species. Uncommon. Mid-mountain forest. Baiyer River. Wild.

166. **Flower Beetle** family Cetonidae. Common. Lowland forest. Long Island. Wild.

167. **Jewel Beetle** *Cyphogastra albertisi*. Uncommon. Lowland rainforest. Brown River near Port Moresby. Captive.

168. **Solitary Bee** *Anthophora* species asleep. Uncommon. Lowland rainforest. Popondetta. Wild.

169. **Orb Weaver** *Argiope* species eating skink. Common. Lowland savannah. Weam. Wild.

170. **Lichen Spider** family Sparassidae. Uncommon. Lowland rainforest. Brown River near Port Moresby. Wild.

171. **Bird-eating Spider** *Selenocosmia crassipes*. Common. Lowland savannah and rainforest. Brown River near Port Moresby. Wild.

172. **Green Jumping Spider** *Mopsus mormon*, male. Common. Lowland savannah. Port Moresby. Wild.

173. **Rhododendron.** *Rhododendron aurigeranum*. Uncommon. Mid-mountain forest. *Drawing*: Vali Herzer.

174, 175. *Bulbophyllum* species. UPNG Orchid Collection.

176. *Bulbophyllum grandiflorum*. UPNG Orchid Collection.

177. **Star Orchid** *Bulbophyllum fractiflexoides*. Uncommon. Mid-mountain forest. Baiyer River. Garden.

178. **Cherry Orchid** *Epiblastus basalis*. High mountain moss forest. Mt Albert Edward. Wild.

179. **Ground Orchid** *Calanthe engleriana*. Common. Mid-mountain forest. Baiyer River. Wild.

180. **Pink Ground Orchid** *Spathoglottis plicatus*. Common. Disturbed areas in lowlands and mid-mountains. Wau. Wild.

181. **Flame Orchid** *Dendrobium flammula*. Uncommon. High mountain moss forest and alpine grassland. Mt Albert Edward. Wild.

182. **Scented Orchid** *Dendrobium anosmum*. Widespread in lowland forest. UPNG Orchid Collection.

183. **Bottlebrush Orchid** *Dendrobium smillieae*. Common. Lowland rainforest. UPNG Orchid Collection.

184. **Shy Orchid** *Dendrobium williamsianum*. Common but restricted distribution in Central District. UPNG Orchid Collection.

185. **Rhododendron** *Rhododendron zoelleri*. Widespread. Mid-mountain and high mountain forest. Baiyer River. Garden.

186. **Macgregor's Rhododendron** *Rhododendron macgregoriae*. Common. Mid-mountain forest and high-mountain moss forest. Baiyer River. Garden.

187. Buds of *Eugenia* species. Common in lowland forest. Brown River near Port Moresby. Wild.

188. **Moss forest Fruit.** Not identified. Uncommon. High mountain forest. Tomba near Mt Hagen. Wild.

189. **Moss forest Fruit.** Not identified. Common. High mountain forest. Tomba near Mt Hagen. Wild.

190. **Flame-of-the-Forest** *Mucuna novaeguineensis*. Common. Lowland forest. Port Moresby. Garden.

191. **Pitcher Plant** *Nepenthes mirabilis*. Uncommon. Poor soils in lowland savannah and forest edge. Weam. Wild.

192. **Sundew** *Byblis liniflora*. Rare. Poor soils of south-western Papuan savannah. Weam. Wild.

193. **Pitcher Plant.** See 191.

194. **Agile Wallaby.** See 16. *Drawing*: Vali Herzer.

195. **Forest Bittern** *Zonerodius heliosylus*. Uncommon. Rivers and streams in mid-mountain forest. *Drawing*: Vali Herzer.

196. **Stylised** *Paradisaea* Bird of Paradise. *Drawing*: Vali Herzer.

Green Tree Python, juvenile. See 77 and back cover.

194

INDEX—COMMON NAMES

INDEX— SCIENTIFIC NAMES